This book belongs to:

這本書屬於:

U0130555

_____

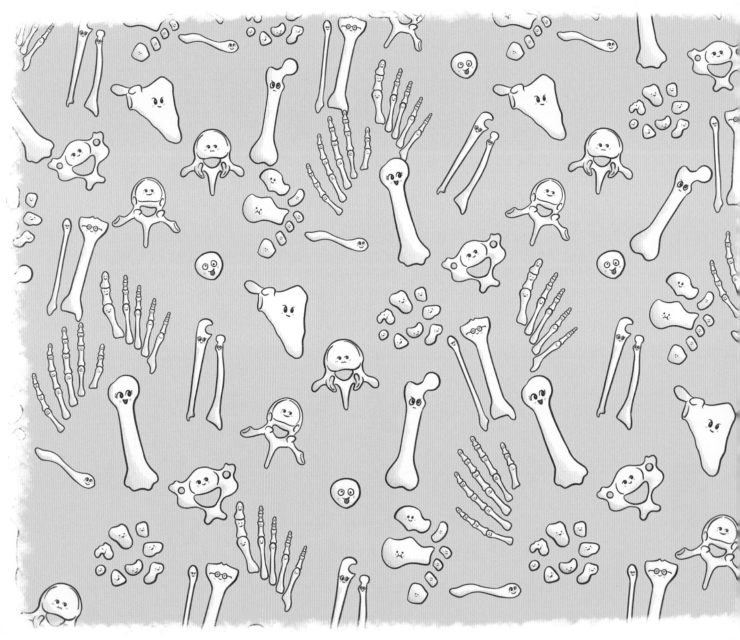

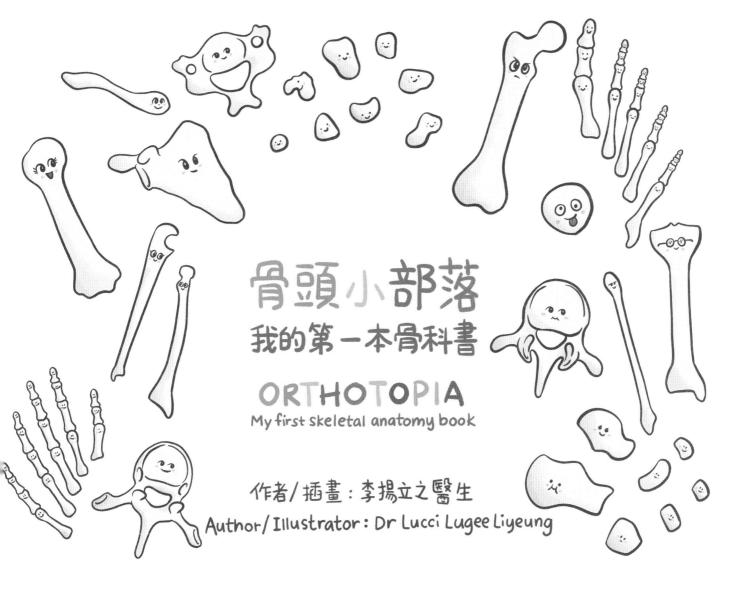

骨頭小部落
我的第一本骨科書

ORTHOTOPIA
My first skeletal anatomy book

作者/插畫：李揚立之醫生
Author/Illustrator：Dr Lucci Lugee Liyeung

我在看小朋友病人的時候
經常會遇到的情況就是
向他們講述受傷的地方在哪裏。
發現要解釋不同的骨頭並不容易喔，
所以乾脆畫一本介紹骨頭的書吧！

It was never easy to explain to young
children where their bony injuries were.
This drove me to publish my own book that
teaches anatomy in a fun and easy way!

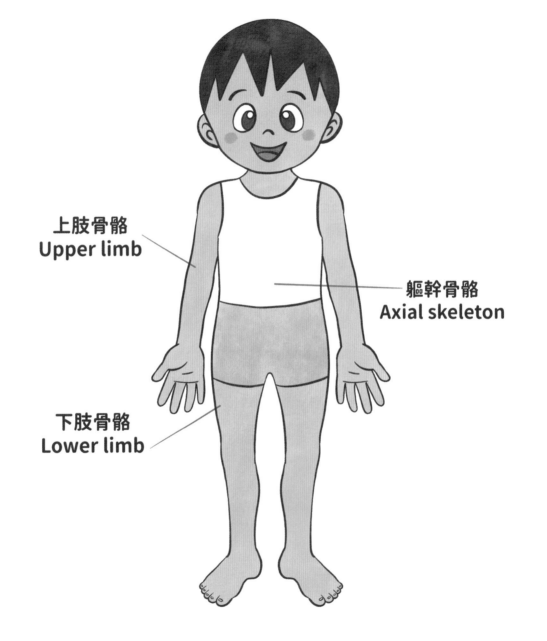

上肢骨骼
Upper limb

軀幹骨骼
Axial skeleton

下肢骨骼
Lower limb

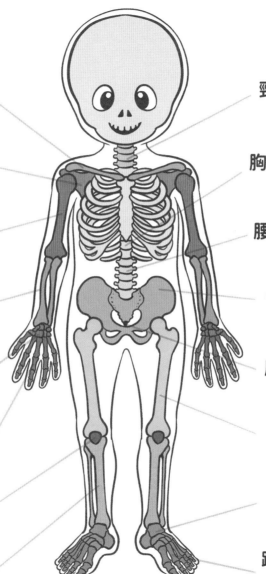

鎖骨 Clavicle — 10

肩胛骨 Scapula — 14

肱骨 Humerus — 18

橈骨與尺骨
Radius and Ulna — 22

腕骨 Carpal bones — 26

掌骨與指骨 Metacarpals
and Phalanges— 30

髕骨 Patella— 44

脛骨與腓骨
Tibia and Fibula — 40

頸椎 Cervical Spine — 58

胸椎 Thoracic Spine — 62

腰椎 Lumbar Spine — 66

盤骨 Pelvis — 70

尾骨 Coccyx — 74

股骨 Femur — 36

跗骨 Tarsal bones — 48

蹠骨與趾骨 Metatarsals
and Phalanges— 52

上肢骨骼
Upper limb

鎖骨
Clavicle

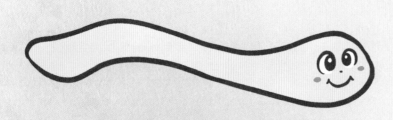

我是鎖骨，是肩膀上方的一塊細長的
S 形骨頭，將手臂與身體主軸連接起來。
I am the clavicle, also known as the
collarbone. I am a long, slender,
S-shaped bone over the shoulder
that connects the arm to the body.

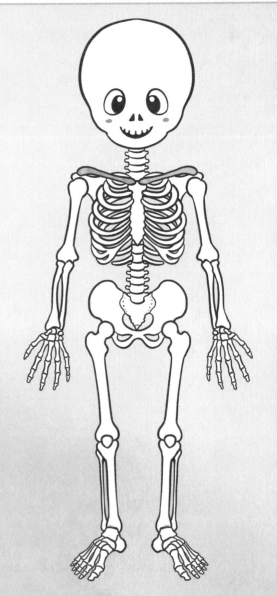

有了鎖骨，你就能牢牢的掛住自己！
With your clavicle, you can hang on tight!

斗毛醫生話你知：
Dr Dumo's fun facts:

貓咪的鎖骨特別短小，這樣可以令牠們輕易地通過狹窄的通道喔！
Cats have tiny clavicles.
This allows them to pass through narrow places easily!

肩胛骨
Scapula

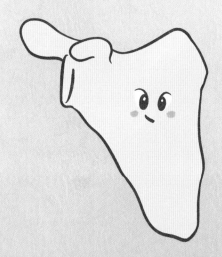

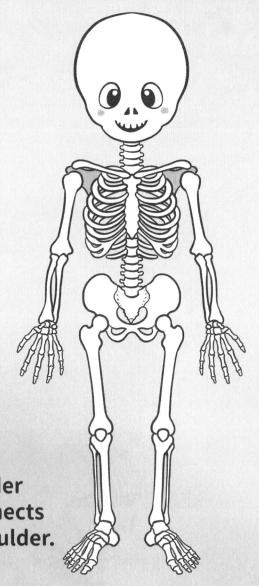

我是肩胛骨，一塊扁平的三角形骨頭，
與鎖骨和肱骨相連，形成肩關節。
I am the scapula, also known as the shoulder
blade. I am a triangular flat bone that connects
the clavicle and humerus, forming the shoulder.

有了肩胛骨，你就能舉手擊掌！
With your scapula, you can give a High Five!

斗毛醫生話你知：
Dr Dumo's fun facts:

在古代，有很多國家的人會在動物的肩胛骨上寫字作占卜用途！
In the ancient times, fortune tellers would paint on animal scapula bones to predict the future.

16

肱骨
Humerus

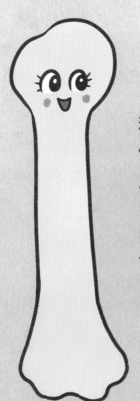

我是肱骨（肱讀音轟），
你們的上臂骨頭，亦是人類
上肢骨頭中最長的一條。
I am the humerus, the
upper arm bone. I am
the longest bone in the
human upper limb!

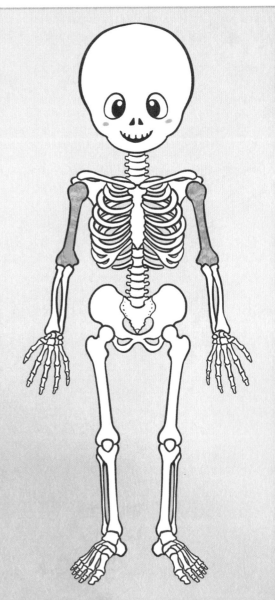

有了肱骨，你就能給人大大的擁抱！
With your humerus, you can give big hugs!

斗毛醫生話你知：
Dr Dumo's fun facts:

當你撞到手肘內側的時候，手會覺得很痺痛。
這是因爲位於肱骨內側的尺神經受到刺激。
You feel tingling pain on your hands when
you hit your "funny bone" at your elbow.
That's because the ulnar nerve, which lies
at the inside of your humerus, gets hit.

橈骨與尺骨
Radius and Ulna

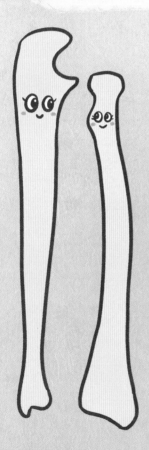

我們是橈骨與尺骨，一起組成你們的前臂骨頭，並分擔移動手臂的工作。
We are the radius and ulna. Together we make up your forearm, and share the function to move your arm.

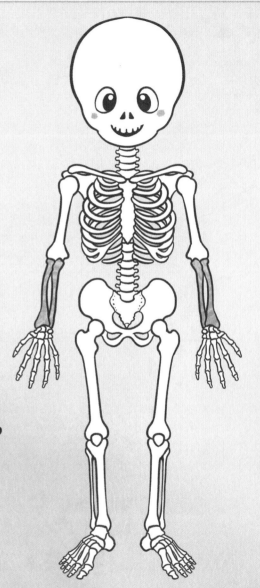

有了橈骨和尺骨，你就能吃自己喜歡的食物！
With your radius and ulna, you can eat the food you love!

斗毛醫生話你知：
Dr Dumo's fun facts:

腕骨
Carpal bones

鈎骨
Hamate

頭狀骨
Capitate

小多角骨
Trapezoid

大多角骨
Trapezium

豌豆骨
Pisiform

三角骨
Triquetrum

月骨
Lunate

手舟骨
Scaphoid

腕骨小隊共有八名骨頭成員位於手腕，
我們讓你的手腕變得靈活而強壯。
The carpal squad has eight bones at the wrist.
We make your wrist flexible and strong.

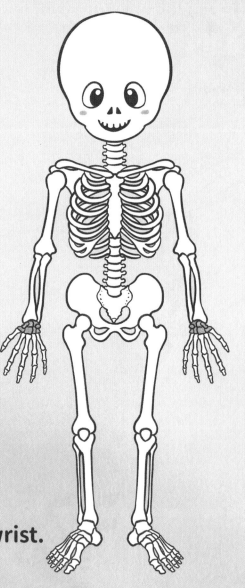

有了手腕骨，你就能把想像的都繪畫出來!
With your carpal bones, you can draw what you imagine!

斗毛醫生話你知：
Dr Dumo's fun facts:

大約每八個右撇子，
就會有一個左撇子！
For every eight right-handers,
there will be one left-hander!

掌骨與指骨
Metacarpals and Phalanges

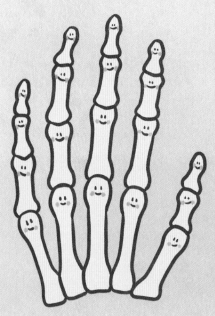

每隻手共有五條掌骨和十四條指骨，組成我們的大拇指、食指、中指、無名指和尾指。

There are 5 metacarpals and 14 phalanges in a hand, forming the thumb, index, middle, ring and little fingers.

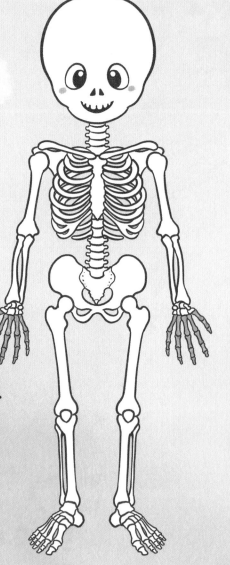

有了掌骨和指骨，你可以告訴人你喜歡他喔！
With your metacarpals and phalanges,
you can show someone that you like them!

斗毛醫生話你知：
Dr Dumo's fun facts:

靈長類如人類的手和其他動物的分別是可相對的拇指，給予我們緊握物件的能力。
What makes us different from other animals is our opposable thumbs, which give us the ability to grab things tight.

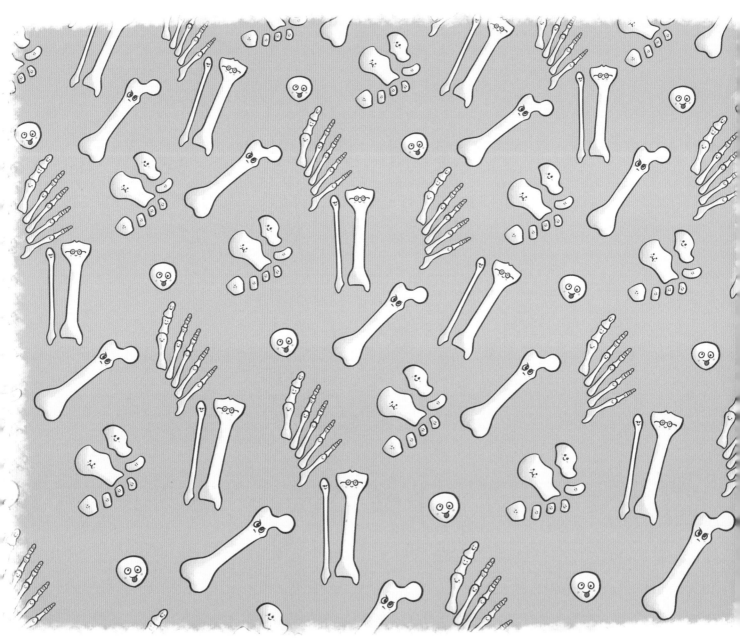

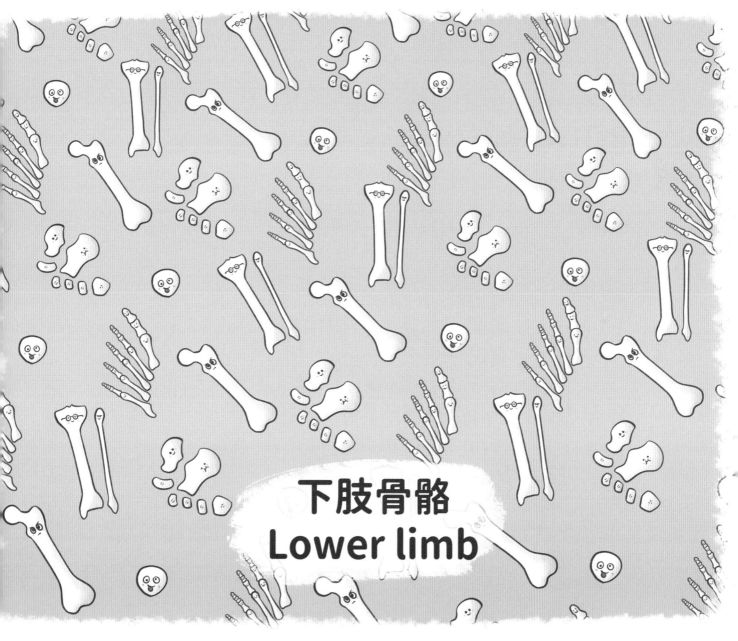

下肢骨骼
Lower limb

股骨
Femur

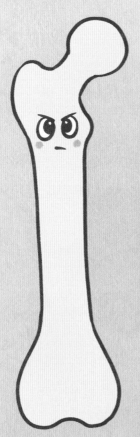

我是股骨，你們的大腿骨頭，亦是你身體中最長的一條骨頭！
I am the femur, the thigh bone. I am the longest bone in your body!

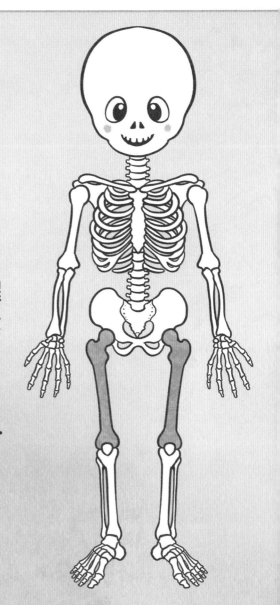

有了股骨，你就能向前跳躍！
With your femur, you can take a big leap forward!

斗毛醫生話你知：
Dr Dumo's fun facts:

股骨是老婆婆受傷中，最常見骨折的骨頭。
The femur is the most commonly fractured bones amongst old grannies.

脛骨與腓骨
Tibia and Fibula

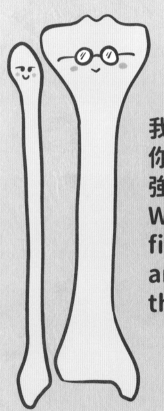

我們是脛骨與腓骨，位於你的小腿。而脛骨是比較強壯的那一條！
We are the tibia and fibula, your shin bone and calf bone. Tibia is the stronger of the two!

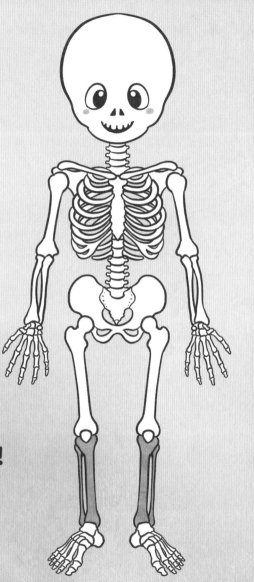

有了脛骨和腓骨，你就能在城市到處走走！
With your tibia and fibula, you can stroll around the city!

斗毛醫生話你知：
Dr Dumo's fun facts:

由羊、鹿、鳥等動物脛骨製成的骨笛
是世界上最古老的樂器之一。
Flutes made of animal tibia bones, such as those of sheep, deers and birds, are one of the oldest musical instruments in the world.

髕骨
Patella

我是髕骨，你的膝蓋骨。
我就是你膝關節前面的那個骨頭。
I am the patella, your kneecap.
I am the bone at the front of
your knee joint.

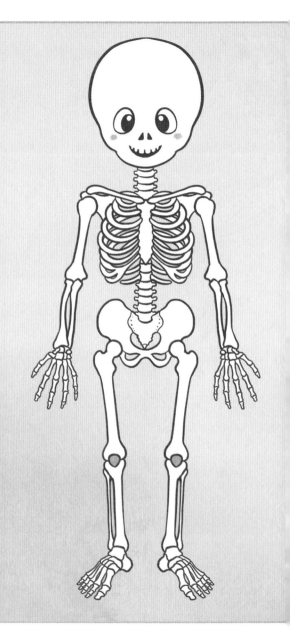

有了髕骨，你就能做各種運動！
With your patella, you can do all kinds of sports!

斗毛醫生話你知：
Dr Dumo's fun facts:

每個人的髕骨都是獨特的！
就如指紋一樣，可作身份識別。
Just like our fingerprints, our
kneecaps are unique, and can
be used for identification!

跗骨
Tarsal bones

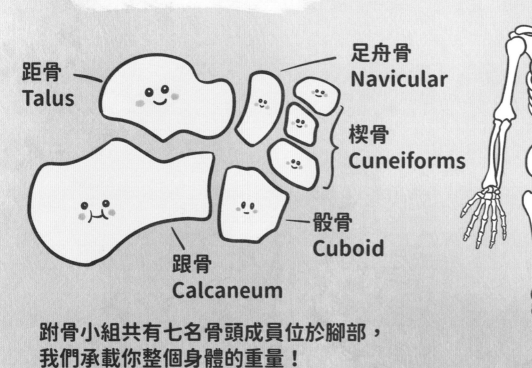

距骨
Talus

足舟骨
Navicular

楔骨
Cuneiforms

骰骨
Cuboid

跟骨
Calcaneum

跗骨小組共有七名骨頭成員位於腳部，
我們承載你整個身體的重量！
The tarsal troops has seven bones in the foot.
We carry the weight of your whole body!

有了跗骨，你就能立正站穩。
With your tarsal bones, you can stand tall and firm!

斗毛醫生話你知：
Dr Dumo's fun facts:

所有嬰兒天生都是扁平足，
足弓要在七八歲才會完成發育的。
All babies are born with flat feet.
The foot arch does not form until
7-8 years old.

蹠骨與趾骨
Metatarsals and Phalanges

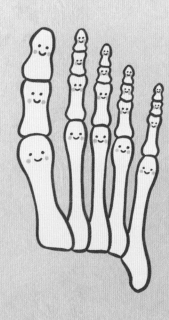

每隻腳共有五條蹠骨和十四條趾骨，組成我們的大拇趾、 二趾、 三趾、 四趾和小趾。
There are 5 metarsals and 14 phalanges in a foot, forming the great, second, middle, forth and little toes.

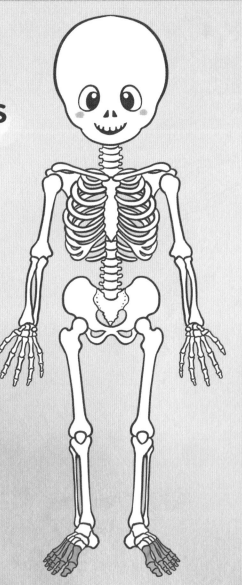

有了蹠骨和趾骨，你就能踮起腳尖觸及得更高！
**With your metatarsals and phalanges,
you can stand tiptoe and reach higher!**

斗毛醫生話你知：
Dr Dumo's fun facts:

在動物世界，蹠骨的數目和腳趾的數目不一定一樣的！
The number of metatarsal is not always the same as the number of toes in the animal kingdom!

54

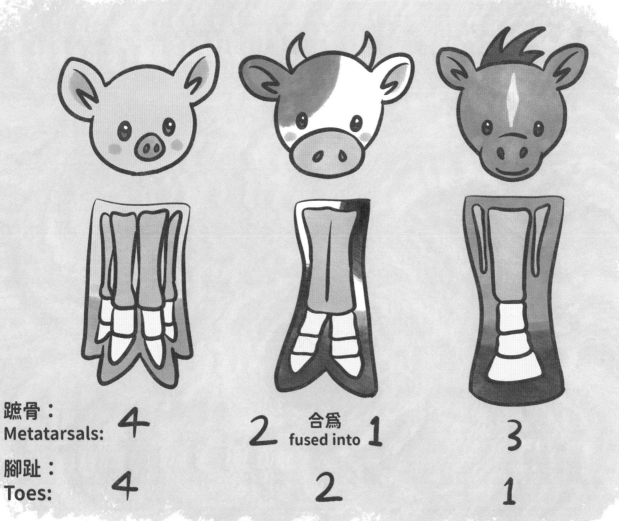

蹠骨：
Metatarsals: 4   2 合為 fused into 1   3

腳趾：
Toes: 4   2   1

軀幹骨骼
Axial skeleton

頸椎
Cervical Spine

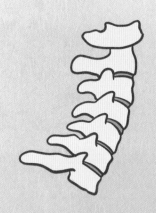 

我們是頸椎，脊椎的前七塊骨頭，
位於頸部區域。
We are the cervical spine, the first
seven bones of your backbone
located at the neck region.

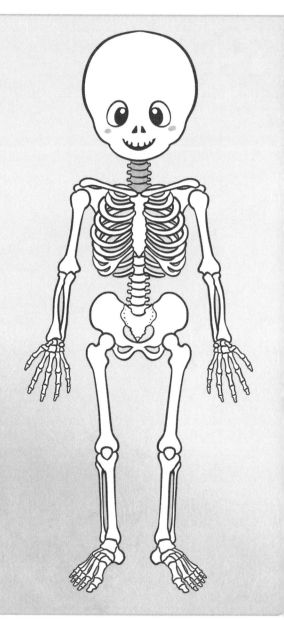

有了頸椎，你就能仰望天上的星星！
With your cervical spine, you can look up at the stars!

斗毛醫生話你知：
Dr Dumo's fun facts:

長頸鹿雖然頸部長長的，但是和我們一樣都只有七顆頸椎骨哦！
Giraffes have such long necks, but they have only seven cervical bones just like us!

胸椎
Thoracic Spine

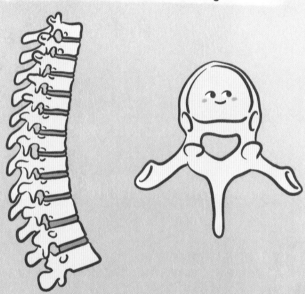

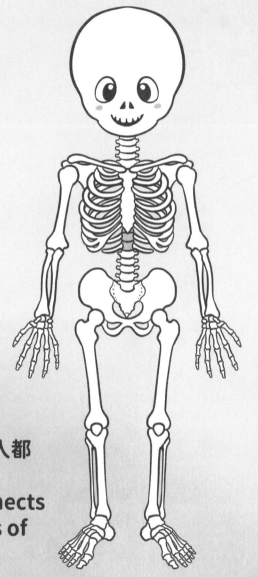

我們是胸椎，每顆都會連接着一對肋骨。多數人都有12對肋骨，但是，有些人有13對肋骨哦！
We are the thoracic spine, each of us connects to a pair of ribs. Most people have 12 pairs of ribs, but some special ones will have 13!

有了胸椎，你就能深呼吸新鮮空氣！
With your thoracic spine, you can take deep breaths of fresh air!

斗毛醫生話你知：
Dr Dumo's fun facts:

劍龍背脊的板片可以用作防禦、
降溫，還有裝飾的用途。
The plates along the back of the stegosaurus were used for self-defence, cooling down, or simply for showing off.

腰椎
Lumbar Spine

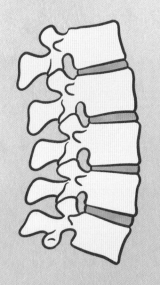
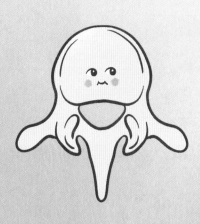
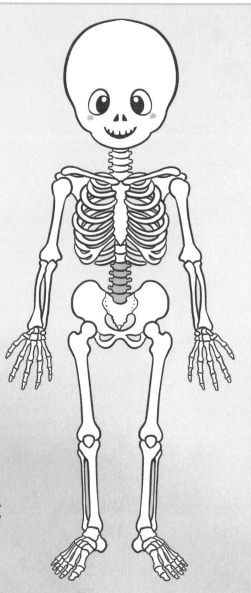

我們是腰椎，位於脊椎的下背部，
一共有五塊骨頭。
We are the lumbar spine, located at
your lower back region.
There are five of us.

有了腰椎，你就能撿起地上的東西。
With your lumbar spine, you can pick up things on the floor!

斗毛醫生話你知：
Dr Dumo's fun facts:

因爲沒有了地心吸力對脊椎的影響，當太空人上了太空便會長高了幾公分。
Without the effects of gravity on their spine, astronauts become a few centimetres taller while in space.

盤骨
Pelvis

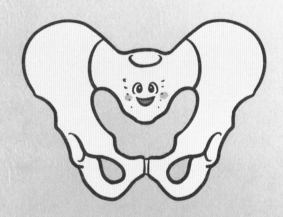

我是盤骨。我連接你的上半身和下肢，並保護你的內臟！
I am the pelvis. I connect your upper body with your lower limbs and protect your internal organs!

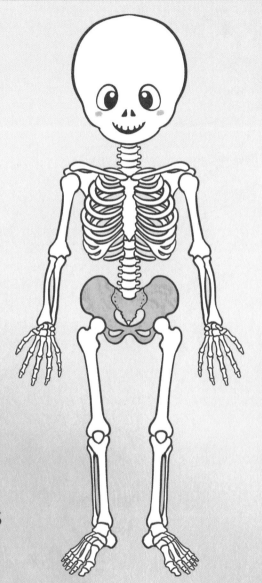

有了盤骨，你就能坐得穩！
With your pelvis, you can sit up stably!

斗毛醫生話你知：
Dr Dumo's fun facts:

鯨魚雖然在水裏生活而沒有下肢，但是牠那個細小的盤骨，就是牠的祖先能在陸地行走的證據。
Whales live underwater and do not have lower limbs. But their tiny pelvis is the evidence that their ancestors had lived on land.

尾骨
Coccyx

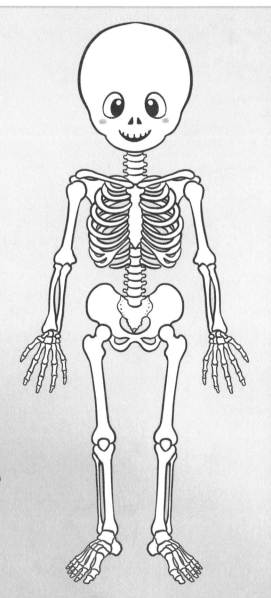

我是尾骨，脊椎底部的最後一塊骨頭。
我是一根細小而彎曲的骨頭。
I am the coccyx, the last bone at the
bottom of your backbone. I am a
small and curved bone, also known
as the tailbone.

咦？尾骨有甚麼作用呢？
Hmm, what's the use of our tailbone?

斗毛醫生話你知：
Dr Dumo's fun facts:

尾骨是人類進化後遺留下來的尾巴。
The coccyx is a remnant of the tail humans used to have before evolution.

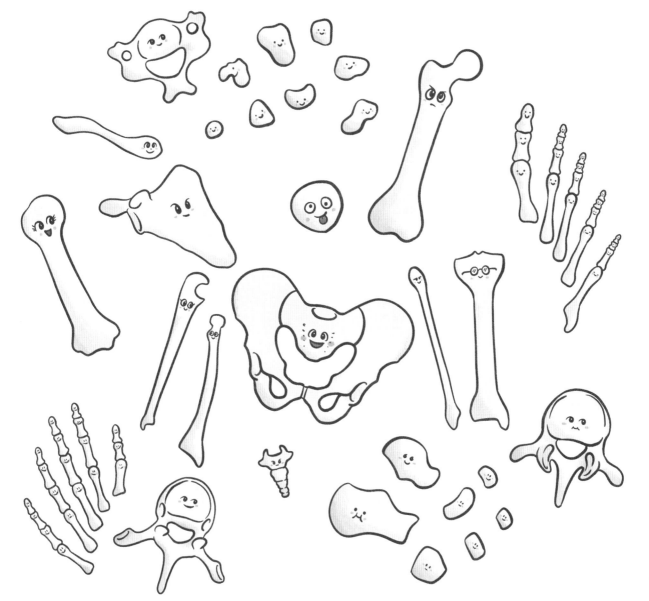

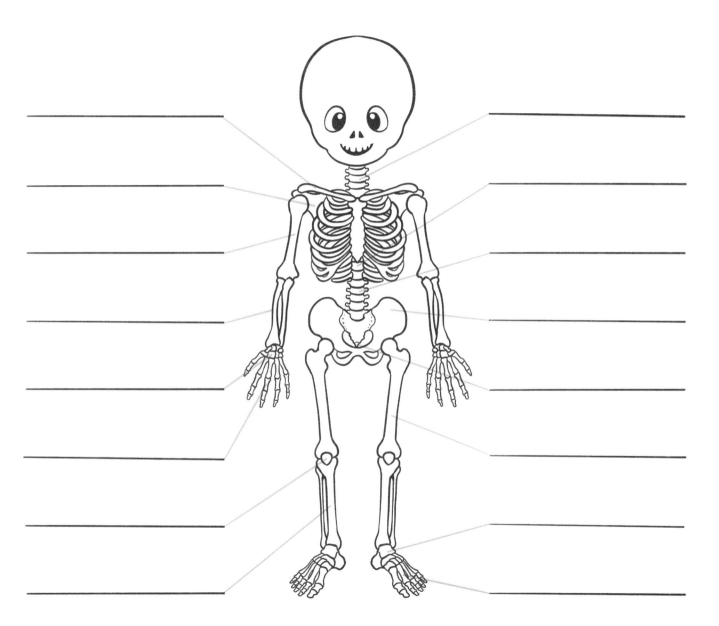

書　　名	骨頭小部落：我的第一本骨科書	
	ORTHOTOPIA: My first skeletal anatomy book	
作　　者	李揚立之	
責任編輯	王穎嫻	
美術編輯	蔡學彰	
出　　版	小天地出版社（天地圖書附屬公司）	
	香港黃竹坑道46號新興工業大廈11樓（總寫字樓）	
	電話：2528 3671　　傳真：2865 2609	
	香港灣仔莊士敦道30號地庫（門市部）	
	電話：2865 0708　　傳真：2861 1541	
印　　刷	亨泰印刷有限公司	
	柴灣利眾街德景工業大廈10字樓	
	電話：2896 3687　　　　傳真：2558 1902	
發　　行	聯合新零售（香港）有限公司	
	香港新界荃灣德士古道220-248號荃灣工業中心16樓	
	電話：2150 2100　　　　傳真：2407 3062	
出版日期	2024年7月初版・香港	

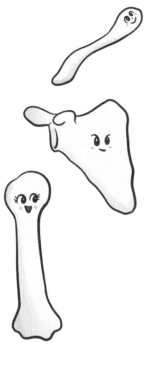
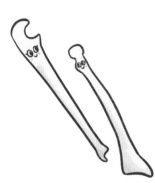

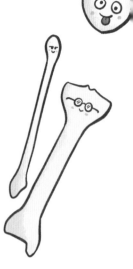

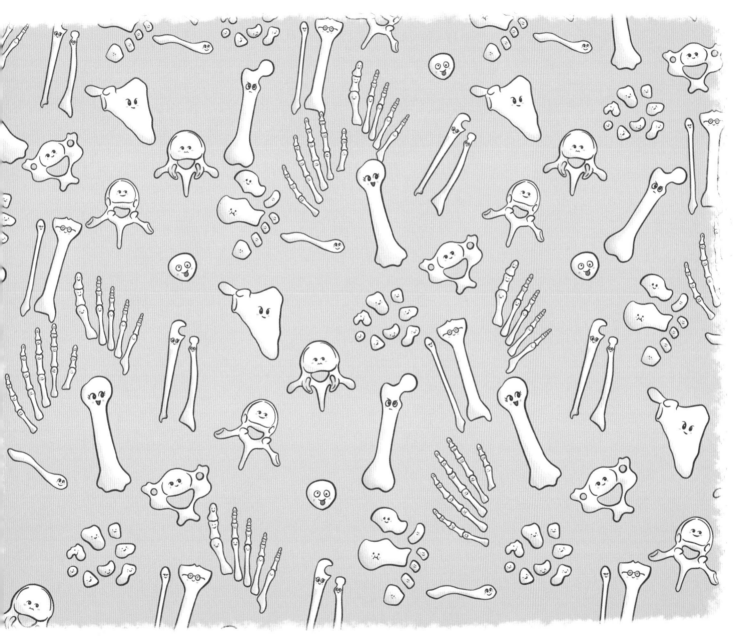

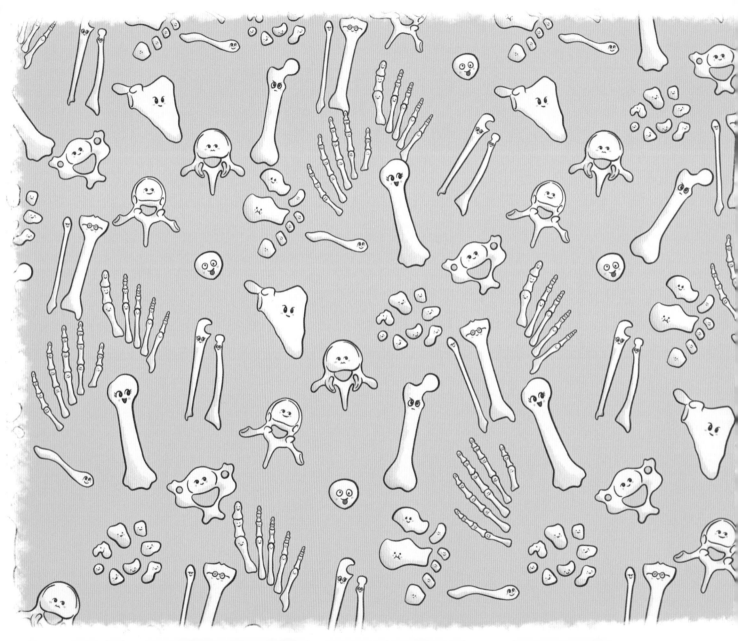

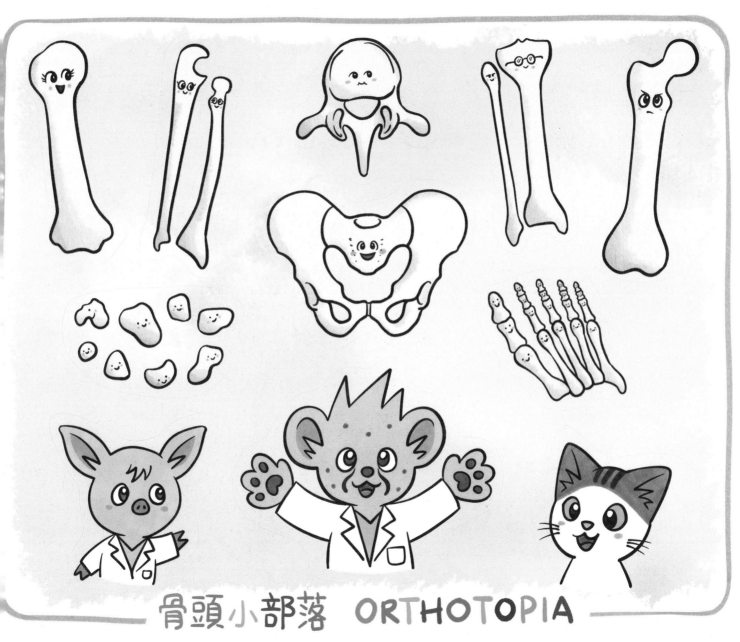

骨頭小部落 ORTHOTOPIA